IMPERIAL
FIRE APPARATUS
1969 THROUGH 1976
PHOTO ARCHIVE

Richard J. Gergel

Iconografix
Photo Archive Series

Iconografix
PO Box 446
Hudson, Wisconsin 54016 USA

Library of Congress Card Number: 2002112571

ISBN 1-58388-091-7

03 04 05 06 07 08 09 5 4 3 2 1

Printed in China

Cover and book design by Shawn Glidden

Copyediting by Suzie Helberg

COVER PHOTO: Shirley, Massachusetts, owned this handsome-looking 1973 Imperial (I-608) Model D-10 Triple Combination Pumper. *Photo courtesy Shirley Fire Department, J.L. Heath photo*

BOOK PROPOSALS

Iconografix is a publishing company specializing in books for transportation enthusiasts. We publish in a number of different areas, including Automobiles, Auto Racing, Buses, Construction Equipment, Emergency Equipment, Farming Equipment, Railroads & Trucks. The Iconografix imprint is constantly growing and expanding into new subject areas.

Authors, editors, and knowledgeable enthusiasts in the field of transportation history are invited to contact the Editorial Department at Iconografix, Inc., PO Box 446, Hudson, WI 54016.

ACKNOWLEDGMENTS

A heartfelt gratitude to the many Firebuffs, Chiefs, Firemen, Amateur and Professional Photographers, Salesmen, and the Waterous Company who contributed the information, research, and photographs to memorialize the History of Imperial Fire Apparatus which allowed the re-construction of the Company's History over a quarter of a century after ceasing business. Original factory photographs are from Ralph Aspling's collection. A special thanks to Ralph Aspling and Skip Stinger for getting me started. Any inaccuracies in the publication of the Company's record are un-intentional.

BIBLIOGRAPHY

American Fire Engines Since 1900, Walter McCall, 1976
Aerial Fire Trucks, Larry Shapiro, 2002

Other helpful sources included Imperial sales brochures and literature and verbal discussions with former Imperial employees.

Aerial photograph of General Administration Offices and Manufacturing facility of Imperial Fire Apparatus, a sales division of Pemberton Fabricators, Inc., an Inductotherm Company at 30 Indel Avenue, Rancocas, New Jersey. *Photo courtesy of Ralph Aspling*

INTRODUCTION

Historically, since their introduction, fire trucks have exercised a particular appeal that fascinates both kids and grown-ups of all ages. The manufacturing of these units has greatly evolved during the past decades and so our history of Imperial fire apparatus has evolved, quite by accident.

The story begins, when on July 5, 1962, Ralph Aspling founded Pemberton Fabricators at the corner of Hanover Street and Route 38 in Pemberton, New Jersey. Ralph, a coppersmith by trade, worked for RCA prior to his entrepreneurial venture with the start up of Pemfab. The company fabricated various metal fabrications, including a substantial volume of copper fabrications for RCA for the broadcast and communications industry and a wide range of metal fabrications for a multitude of companies including Inductotherm Industries. Pemberton Fabricators was acquired by Inductotherm in May of 1967 and was moved to a new and larger facility in the spring of 1968 on Indel Avenue in Rancocas, New Jersey (a community located approximately 17 miles northeast of Philadelphia), within the Corporate Complex of Inductotherm Companies.

As fate would have it, one day the CEO of Inductotherm asked Ralph to help one of his friends in fabricating a trailer with a fire fighting capability package. While completing this task, Ralph met Jim Partridge while buying parts at Hale Pump Company in Conshohocken, Pennsylvania, who had numerous years serving the needs of the fire industry and who was currently working for Hahn Motor Company in Hamburg, Pensylvania. The two hit it off and shortly thereafter, in 1969, Jim Partridge was hired as Sales Manager of Fire Apparatus by Pemberton Fabricators to develop the fire apparatus division which he would name "Imperial Fire Apparatus."

The first truck to be manufactured was a mini pumper for the Linfield Volunteer Fire Department in Pennsylvania. The first custom Imperial rig was delivered to the Goodwill Fire Company Number 11, in Pemberton, New Jersey in 1970. Vernon Ney was responsible for the design of the first diesel powered custom engine, which was sold in 1997 and is still in service at the Eoline Volunteer Fire Department, Centreville, Alabama today. Imperial's first trade advertisement appeared in an industry publication in January of 1971. The ad featured Pemberton's custom engine (I-505), as well as a light and brush apparatus for Gladwyne,

Pennsylvania (I-501), a 2,000-gallon tanker for Elverson, Pennsylvania (I-502) and a mini pumper for Linfield, Pennsylvania (I-500).

New orders and sales were good and within a few months orders were secured for a snorkel, a midship-mounted and a rear-mounted aerial ladder. Time would show that the results on units from a profit and loss standpoint were very disappointing. In mid 1973 Ray Wells, the Vice President of Sales for Pemberton Fabricators, wrote a report to the Board of Directors of the company. Wells recommended a change of direction devoting the company's efforts from the sales of just chassis toward the sales of a wide range of Original Equipment Manufacturers (OEMs), which was approved. Ray Wells assumed the new additional responsibility for product sales of "Chassis by Pemfab," which would later evolve. Partridge, now the Vice President of Fire Apparatus Sales, left the company in the fall of 1973, and Ralph Aspling, the President, followed in December. Contrary to incorrect reports of a bankruptcy filing or sale, the company simply stopped the sales of completed apparatus and concentrated on the sales of fire chassis only. All sales in process were honored and in September 1973 the first advertisement appeared, "Specify Imperial Chassis" in *Fire Engineering* Magazine.

Imperial fire chassis were being sold to Fire Trucks, Inc., Pierce Manufacturing, Pierreville Fire Trucks, and Thibault, to name a few. The company now was promoting "Chassis by Pemfab," which continued until 1985 when the identity was revised to "Pemfab Trucks." Pemfab industry firsts included being the first manufacturer to install a tilt and telescopic steering column and the first to produce a 10-man full-tilt four-door cab, both accomplished in 1985. In August of 1997, the Company's technical data package records, including customer lists, engineering drawings and bill of materials, etc., excluding the name Pemfab Trucks, was sold to Fire Cab, a division of CECO/Taylor Company, a subsidiary of Chambersburg Engineering. Fire Cab ceased their fire truck operations in July of 2000. KME (Kovatch Mobile Equipment) of Nesquehoning, Pennsylvania purchased the inventory at auction in March of 2001, concluding the story of heavy duty, tough, reliable Imperial and Pemfab Trucks.

Little is known about this Fire Fighting Trailer that was manufactured by Pemberton Fabricators (circa 1969) except for the fact that it influenced Company Management to embrace the fire fighting industry. Shortly thereafter Imperial Fire Apparatus was formed as a sales division of Pemberton Fabricators. *Photo courtesy of Ralph Aspling*

Spec. No. I-500 Linfield, PA. - The Linfield Volunteer Fire Company Number 1 is credited with receiving the first of the Imperial units, a mini pumper in mid 1970. The rig was built on a 4 x 4 Dodge W-500 chassis powered with a 177 HP gas engine and had a 7,000-pound front axle, a 17,000-pound rear axle, a 500-gpm pump and a water tank of 500 gallons. The unit was sold to Lawn Fire Company in Lawn, Pennsylvania in August 1991 and, after being repainted lime-yellow and white, was placed in service in June 1992 where it remains in service today. After 30 years of service, there are only 19,011 miles registered on its odometer. *Photo courtesy of Matt Beare*

Spec. No. I-500 Lawn, PA. - Lawn Fire Department deserves another view. In the 10-month period between purchasing and placing the unit in service, the following was added: a bumper brush guard, a roof-mounted bar light, a portable generator, a Honda 4,000-watt and two 500-watt portable flood lights, a hose reel for a garden hose, an aluminum booster tank with lifetime guarantee, a hose bed for 350 feet of 3-inch rubber hose, and two 10-foot lengths of 4-inch soft suction hose. The unit was also repainted lime-yellow and two-toned with a white hood and roof. The 33-year-old unit was sold to a private individual in September 2002. *Photo courtesy of Matt Beare*

Spec. No. I-502 Elverson, PA. - The Elverson Fire Company Number 1 accepted their Imperial tanker in July 1970. The cab-over Ford C-900 3-man tilt cab chassis GVWR was 38,000 pounds with a 15,000-pound front axle and a 23,000-pound rear axle. It had a 534 engine, a manual Spicer transmission on a 175-inch wheelbase, and a CA dimension of 146 inches. The tanker fire package consisted of a 250-gpm power take-off pump with a 2,000-gallon booster tank. Fenderettes and a rub rail were part of the first tanker design. The unit served the department until it was taken out of service on August 11, 1995, and sold to Sherwood Fire Department in Arkansas for $4,100. *Photo courtesy of Bob Gebhardt*

Spec. No. I-505 Pemberton, NJ. - Goodwill Fire Company Number 1 placed the first Imperial custom pumper in service, a Model D-10 in September 1970. A 6-71N, 265 HP Detroit Diesel engine powers this rig built on a 166-inch wheelbase with a manual Spicer transmission, a 1,000-gpm pump and a 500-gallon booster tank. It served in the department until April 1997, when it was sold to a broker who resold it to Brent Volunteer Fire Department in Alabama. It was resold to Eoline Volunteer Department, a neighboring community in Centreville, Alabama in August 1997, where it remains in service today. *Photo courtesy of Ralph Aspling*

Spec. No. I-507 Elsmere, DE. - Elsmere Fire Company Number 1 took delivery of the first Imperial custom rescue unit in October 1970. The 18-foot walk-in body was sub-contracted to Providence Body Company, which discontinued manufacturing rescue bodies for the fire service in 1974. The 35,000-pound GVWR chassis with a 180-inch wheelbase had a 235 HP diesel motor, automatic transmission, and ample storage for men and rescue equipment. It was sold to Pillow Volunteer Fire Department of Pennsylvania in the early '90s and resold to Gretna Volunteer Fire Department in Virginia in June 2000. *Photo courtesy of Steelman Hiles*

Spec. No. R-508 Haverford, PA. - The Mano Fire Department was the original owner of the Imperial rescue unit in 1970. The 84-inch CA commercial Dodge W-500 4 x 4 is powered with a 318 HP, V-8 gas engine, 4-speed transmission and has mud and snow tires on all wheels. Extra-value features include a 15-kw gas generator, a manual cable reel, six 1,500-watt telescoping lights, and a squad bench seat inside the rear body. The unit has had a series of owners, sold first to Honey Brook Fire Company in June 1991, again to the Gibraltar Fire Department in Birdsboro, Pennsylvania and again in July 2000 to the York New Salem Fire Company Number 1. *Photo courtesy of Todd J. Lincoln*

Spec. No. I-509 Leighton, PA. - Leighton Fire Company Number 1 received the Imperial brush truck on November 25, 1970. The Dodge W-300 four-wheel drive unit had a power take-off pump capable of pumping 250 gpm supported by a 300-gallon booster tank. A rear-facing seat for three firemen was provided at the rear of the hose body. Ground sweep nozzles were provided along with a deluge gun, and dual booster reels were provided to attack miscellaneous brush fires. The unit was sold in the spring of 1989 to Brown Township Volunteer Fire Company in Cedar Run, Pennsylvania and again by a Ford Agency in Jersey Shore, Pennsylvania to an unidentified source. *Photo courtesy of Steven Ebert*

Spec. No. I-510 Highland, NJ. - Applegarth Fire Company took delivery of their 35,000-pound GVWR chassis, Model D-10 in November 1971. The Detroit Diesel 6-71N in-line was coupled with a Spicer 6852, 5-speed manual transmission with a Waterous 1,000-gpm single-stage CMBX pump and a 750-gallon booster tank. The apparatus had a wheelbase of 178 inches, which became the standard for Imperial (with only a few exceptions). The unit was sold to the Laurel County Fire Department of London, Kentucky in March 1989. The booster tank was replaced with a 1,000-gallon water tank and the unit was repainted in an all-white finish. *Photo courtesy of Ralph Aspling*

Imperial's first advertisement appeared in the January 1971 issue of *Fire Engineering*, a popular trade magazine for the fire industry. Shown is the first custom triple-combination pumper (I-505) delivered to Pemberton, New Jersey in September 1970, a light and brush truck (I-501) for Gladwyne, Pennsylvania, a tanker (I-502) manufactured for Elverson, Pennsylvania, and the first apparatus, a brush truck (I-500) made for Lawn, Pennsylvania. The advertisement appeared again in the April 1971 issue.

Spec. No. I-512 Deptford, NJ. - The Deptford Fire District purchased four custom Imperial Model D-12, 1,250-gpm triple combination pumpers. The units (from left to right) are: I-512B, Almonesson Lake Fire Department; I-512D, Union Volunteer Fire Department, Blackwood Terrace; I-512A, Tacoma Fire Department; I-512C, New Sharon Fire Department. On multiple orders, an alphabet letter was added to the specification number to identify the customer's unit. This order was very important to the new sales division as it firmly established their identity in the fire truck industry. *Photo courtesy of Ralph Aspling*

Spec. No. I-512A Deptford Township, NJ, Tacoma-Westville Grove Volunteer Fire Company - The original unit, delivered in April 1971 (right), had a white finish and was repainted at a later date with a red lower half while the cab's upper half remained white (below). The Model D-12 had a 178-inch wheelbase, 8V-71N, 350 HP motor, Spicer transmission, 1,250-gpm pump, Waterous CMBX, a 750-gallon water tank, and a transverse locker compartment ahead of the pump which became a standard. The unit was sold in May 1998 to the Coastal Eagle Point

Refinery in Westville, New Jersey and remains in service today. *Photos courtesy of Skip Stinger*

17

Spec. No. I-512B Deptford Township, NJ. - Almonesson Lake Fire Department placed their unit in service in April 1971. One might suppose that all the rigs were finished alike, but this fire department specified "school bus yellow." The cab roof and canopy were repainted adding a white top. This unit was powered similar to its sister company but had additional foam capability with a deluge gun and a 100-gallon foam reservoir. The 5-man, 84-inch-wide cab forward supplier was Truck Cab. The unit's ownership was transferred to the Coastal Eagle Point Refinery, Westville, New Jersey in May 1998 and later exported to one of their refineries in Aruba. *Photo courtesy of Scott Mattson*

Spec. No. I-512C Deptford Township, NJ. - New Sharon Volunteer Fire Department received this Model D-12 in April 1971. The traditional all-red fire truck had a white roof added at a later date. Dissimilar to the previous two units, this apparatus' power train included a Detroit Diesel 8V-71N, 350 HP engine, and a Spicer 5-speed syncromesh transmission. The midship pump, a Waterous CMBX 1,250-gpm two-stage centrifugal pump, was complimented with a 750-gallon water tank. The unit remained in service until May 1998 when the apparatus was sold to the Lawn Fire Company in Glassboro, New Jersey. *Photo courtesy of Scott Mattson*

Spec. No. I-512D Deptford Township, NJ. - Union Volunteer Fire Department, Blackwood Terrace took delivery of this Model D-10 in April 1971. The 178-inch wheelbase allowed for the transverse locker compartment, now an Imperial standard. The original unit was red with a white canopy roof painted at a later date. Warning equipment included a Federal Q2B mechanical siren on the left-hand bumper extension and an electronic siren and public address system on the right-hand bumper extension. The unit was sold in May 1998 to the Coastal Eagle Point Refinery, Westville, New Jersey and later exported to one of their refineries in Aruba. *Photo Courtesy of Scott Mattson*

Spec. No. I-514A Bellmawr, NJ. - Bellmawr Fire and Rescue Number 1 received the model D-12 with an articulating 55-foot snorkel device in November 1971. The pumper-snorkel was the only one to be delivered by Imperial. The 6-71N motor had a rating of 308 HP. Its 6852 transmission, 1,250-gpm pump and 500-gallon water tank all fit nicely on the 178-inch wheelbase. The low profile cab and body was painted white. It was sold to Hillbilly Fire Apparatus, a broker who resold the apparatus on January 27, 1994, to the Southern Stonc County Fire Protection District in Reed Springs, Missouri. It was the second of four pre-owned Imperials to be purchased. *Photo courtesy of Dave Ehrman*

Spec. No. I-514B Bellmawr, NJ. - Bellmawr Park Volunteer Fire Department received this unique low profile 5-man 84-inch-wide cab forward Model D-12 in November 1971. It used the 6-71N, 308 HP (a 265 HP engine was the standard), and the standard transmission was a Spicer 6852 5-speed. The fire department "red" apparatus was equipped with a midship pump, a 1,250-gpm pump, and a 500-gallon tank. Note the boot and coat rails above the wheelhouse compartments—rarely seen on today's apparatus. The rig remained in the department until October 2000 when it was sold to Warwick Woods Campgrounds in St. Peters, Pennsylvania. *Photo courtesy of Scott Mattson*

Spec. No. I-515 Narbeth, PA. - Narbeth Fire Company placed this Model D-4-S custom rear-mounted Grove aerial in service in February 1972. The 48,000-pound GVWR with a 14,000-pound front axle and 34,000-pound rear axle was powered with a Detroit Diesel 8V-71 engine and an Allison HT-70 automatic transmission. A 4-section 100-foot Grove aerial was shipped to Imperial on June 25, 1971. Ladder Towers, Inc. acquired the ladder division in 1972 and their first ladder, a 4-section ladder built for Howe, was exported to Guam and was delivered on March 12, 1973. The unit sold for salvaging on October 14, 1999, to Bernie's Enterprises, Sellersville, Pennsylvania. *Photo courtesy of Ralph Aspling*

Spec. No. I-516 Clifton Heights, PA. - Clifton Heights Volunteer Fire Department received the first Imperial 100-foot midship-mounted Grove aerial ladder Model A-4-S in April 1971. It featured a 38,000-pound chassis, 220-inch wheelbase, 265 HP motor and a manual 5-speed transmission. Reyco spring suspension was supplied. The two-color red with white roof aerial unit was in service until it was sold to the Holmes Fire Company, in Pennsylvania in August 1995 where it served the community until September 11, 2001 when the ladder was rendered inoperable. The rig was sold on "Ebay" for $5,400 and delivered to an individual from West Virginia in December 2001 for repair and resale. *Photo courtesy of Robert Orner*

Spec. No. I-517 Rockledge, PA. - Rockledge Volunteer Fire Company took delivery of this Model D-10 in January 1971. In addition to its fire fighting duties, the company is proud of their engine and displays it in community affairs and parades. The truck has a wheelbase of 178 inches, a 1,000-gpm pump and a 500-gallon tank. A 12-inch chrome-plated fire bell added to the pride of this department. Powered by a 6-71N motor with a manual 6852 syncromesh transmission, it served the community well for over 27 years and in October 1998 it was sold to the North Rural Fire District in Yellowstone, Montana. *Photo courtesy of Ralph Aspling*

Spec. No. I-519 Bethlehem, PA. - The Hecktown Volunteer Fire Company Number 1 received their Ford C-900 Rescue apparatus in the last half of 1971. The 3-man tilt cab was on a 153-inch wheelbase, 27,500-pound GVWR with a 9,000-pound front axle and an 18,500-pound rear axle. The walk-in rear rescue box was approximately 15 feet, with large double door compartments ahead of the rear wheels and a large single door aft of the rear axle. A 20,000-pound electric winch was installed behind the front bumper extension. A contract price of $19,807 was signed on November 23, 1970. The unit's whereabouts is unknown. *Photo courtesy of Dave Ehrman*

Spec. No. I-520 Green Lane, PA. - Green Lane Fire Company placed their Model D-10 in service in July 1971. Features of this all-white rig include a 36,000-pound GVWR, 166-inch chassis, a Detroit Diesel 8V-71N engine, an Allison HT-70 automatic transmission, Timken-Rockwell axles—FF-901 (12,000 pound) front, R-140 (24,000 pound) rear—and 10.00 x 20-12 ply tires with Budd disc wheels. Other features include a Waterous CMBX 1,000-gpm two-stage pump, a 500-gallon water tank, Ross HF-64 steering gear, Vickers VTM-42 power steering pump, Farr air cleaner, and a Leece Neville 125-amp alternator. The rig was sold in June 2001 to Sellersville, Pennsylvania. *Photo courtesy of Ralph Aspling*

Spec. No. I-521 Mount Bethel, PA. - Mount Bethel Fire Department received this brush truck in March 1971 and it is still in service today. The chassis is an International Loadstar 1700 series with four-wheel drive capabilities, 139-inch wheelbase, and a front-mounted winch. The pump is a Waterous CPK-3 booster pump with draft capability from the side and front suction, a 400-gallon water tank, and dual-mounted booster reels. Because of its off-road maneuverability, the body compartmentation was constructed from tread plate, which was an easy surface to maintain. The apparatus cab and chassis and upper body had a school bus yellow finish. *Photo courtesy of Dave Bowen*

Spec. No. I-523 Indian Head, MD. - Potomac Heights Volunteer Fire Department & Rescue Squad placed this commercial tanker in service in July 1971. The 32,000-pound GVWR, GMC-CE00213 chassis with a wheelbase of 189 inches and a cab-to-axle dimension of 124 inches accommodated the 427 V-8 gas engine, a Waterous (CPK-3) 400-gpm pump and a 1,500-gallon water tank. Ample compartment space was provided; note the dual double-door compartments ahead of the rear wheels. The apparatus was finished in red and had a white top. In May 1993 the unit was sold to the Macedonia Volunteer Fire Department, in Gaffney, South Carolina. *Photo courtesy of Bob Collins*

Spec. No. I-524 Braintree, MA. - Braintree Fire Department took delivery of the Model D-10 on July 14, 1971. The 35,000-pound chassis had a 265 HP Detroit Diesel engine, 5-speed Spicer syncromesh transmission, 178-inch wheelbase, a midship–mounted Waterous single-stage 1,000-gpm pump, and a standard 500-gallon tank. Continued improvements included the dual bezel red 6-inch sealed beam warning lights that were mounted on the front of the cab. Finish on the apparatus was a red lower half while the cab and canopy sported a white roof. Truck features included dual electric reels, deluge gun, and transverse cross lays. The unit was salvaged for parts in October 1998. *Photo courtesy of Dick Adelman*

Spec. No. I-525 Chews Landing, NJ. - The Chews Landing Fire Department placed their 1971 Ford F-250 Imperial 4 x 4 brush truck in service in April 1971. The chassis had an extended front bumper for an 8,000-pound electric winch, one electric hose reel with 200 feet of 3/4-inch hose and a 250-gallon water tank. A Hale 25FZZ-B23 (250-gpm) pump was piped to the tank, hose reel and the Akron 509 deck gun. Chews Landing supplied the chassis which cost $10,303 and the Imperial's fire body and equipment cost the District $5,923. On August 12, 1992, the unit was sold to Jon's Truck and Equipment Sales in Rogersville, Missouri for $3,360. *Photo courtesy of Dave Ehrman*

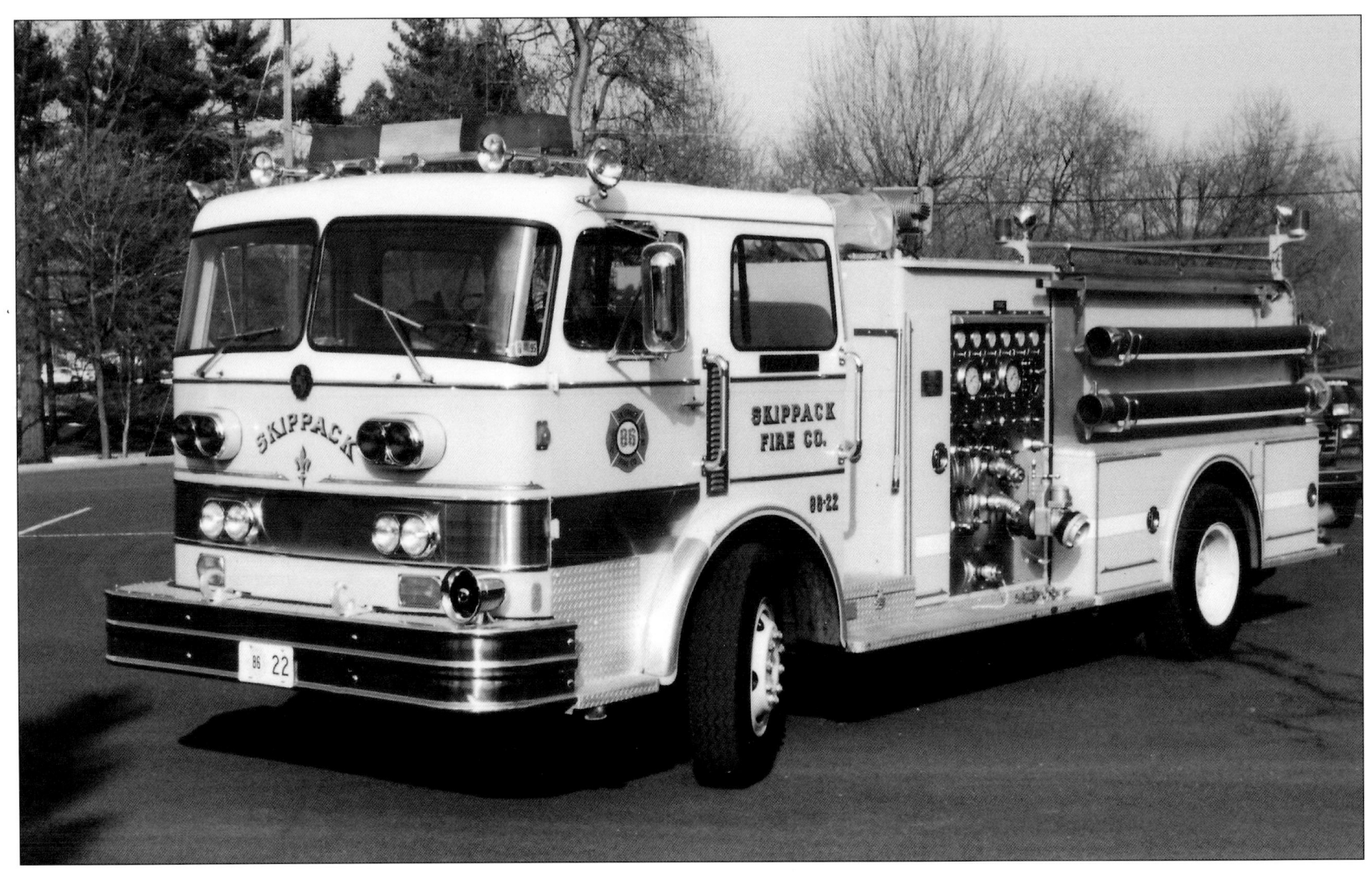

Spec. No. I-526 Skippack, PA. - Skippack Fire Company received their Model D-10 in November 1971. The 178-inch wheelbase, 35,000-pound GVWR chassis was powered by a Waukesha F817G, 308 HP gas engine, Timken FF-901 (12,000 pound) front axle, R-140 (23,000 pound) rear axle, Spicer 6852 transmission, air brakes and a two-stage CMBX 1,000-gpm Waterous pump. The contract price, signed on February 9, 1971, was for $34,320. Value-added features included a 750-gallon tank, 4,000-watt generator, and electric hose reels-rear side compartment mounted. The unit was sold in July 2000 to the Montgomery Fire Academy in Conshohocken, Pennsylvania. *Photo courtesy of Todd Lincoln*

Spec. No. 1-529 Wheaton, MD. - Wheaton Volunteer Rescue Squad placed this 35,000-pound GVWR Imperial 2-man cab chassis, 188-inch wheelbase in service in mid 1972 with a 14-foot, 6-inch walk-in Providence body. Features included a Detroit Diesel engine with an automatic transmission, 15-ton rear winch, two 12-kw diesel generators, a six bottle cascade system, A-frame assembly, two 150-foot and two 75-foot electric reels, plus air bags, rescue tool, five air packs, six spare bottles, smoke ejectors, and hydraulic jacks. Originally this was an all-white unit that later had a red and blue stripe motif added. The rig was taken on trade with a purchase of a new Ranger/Pemfab unit. *Photo courtesy of Skip Stinger*

Spec. No. I-531 Port Norris, NJ. - Port Norris Fire Department received their Model D-10 in July 1971. Originally purchased by Commercial Township Fire District Number 1, the 1,000-gpm pumper (CMB-1) with a 1,000-gallon tank was painted red. This piece received a Detroit Diesel 6-71N motor, a Spicer 6852 transmission with a GVWR of 35,000 lbs., standard body compartmentation and a painted locker compartment door. Guide arrow turn signals were mounted below the headlamps. Additional warning equipment included the 12-inch chrome-plated bell on the right-hand bumper extension and an electronic siren on the left-hand side. *Photo courtesy of Ralph Aspling*

Spec. No. I-531 Port Norris, NJ. - Yes, it's the same rig! The apparatus underwent major re-manufacturing in 1990 by Grumman Emergency Products (which ceased operation in September 1992). The cab was re-built into a four-door fully enclosed model to meet the day's safety standards. A new body was installed complete with dual over-the-wheelhouse compartments, and every detail was addressed right down to the new front square headlamps, emergency and warning lights. The apparatus was repainted with pewter top and a maroon lower half. The apparatus has new life and will serve the department for years to come. *Photo courtesy of W. L. Hollinger*

Spec. No. I-532 Hellam, PA. - Hellam Fire Company received the Model D-10 in December 1971. The pump is rated at 1,000-gpm. The 35,000-pound GVWR chassis is powered with a Detroit Diesel 6-71N motor and has a 5-speed syncromesh transmission. The T-shaped booster tank is a 1,000-gallon tank manufactured from formed 10-gauge hi-tensile all-welded steel construction. Tank interior was coated with two coats of Zincolate. Three removable covers allowed for future cleaning and/or repainting. The tank was provided with a 4-inch over-flow, pump to tank line, 1 1/2 inches, the tank to pump line 2 1/2 inches. The rig serves the department today. *Photo courtesy of Dave Ehrman*

Spec. No. I-533 Throop, PA. - Throop Fire Department E-27 accepted this Ford C-900 266 HP commercial pumper in the first quarter of 1972. It's equipped with a 750-gpm midship pump (CMBX) and a 750-gallon water tank. Covered reel compartment was provided over the pump enclosure for the dual electric reels. The fire body was constructed from formed 12-gauge cold rolled steel and welded to the lower round corner compartment assembly. It featured two generous body compartments—the rear interconnecting with the rear body with tread plate access doors. It remained in service until May 2000 when the unit was sold to a salvage yard in Dunmore, Pennsylvania. *Photo courtesy of Dave Ehrman*

Spec. No. I-534 Townsend, DE. - Townsend Fire Company, Inc. took delivery of the Model D-10 early in 1972. This rig features custom pumpers, front and rear torsion bar suspension, full-time power steering (Ross HF-64), heat treated frames (1,000,000 pounds tensile strength), full air brakes both for service and parking, full width Cincinnati Cabs (84 inches), fiberglass cab roof liner, rub rails and fenderettes, 26 deep compartments, and double pan doors. Powered with a Waukesha F-817G gas engine motor (325 HP) and a 6852 transmission, the 35,000-pound GVWR rig was sold in the 1990s to Beech Island, South Carolina. No records are available. *Photo courtesy of Skip Stinger*

Spec. No. 1-535 Red Bank, NJ. - River Plaza Hose Company Number 1 received their Imperial Model D-12 pumper in February 1972. This rig is powered with a Detroit Diesel 8V-71 (350 HP) engine and syncromesh transmission. The two-stage Waterous CMBX pump is connected to a 500-gallon water tank. The unit is now painted white over red in lieu of the original red finish. The triple combination pumper was sold in June 2000 to Forest Lake Volunteer Fire Company in Forest Lake, whose post office box is located in Montrose, Pennsylvania. Chief Fred Capotosto "expects to run the unit for an additional 20 years," further testimony to Imperial's quality. *Photo courtesy of Ralph Aspling*

Spec. No. I-536 Langhorne, PA. - Parkland Fire Company Number 1 placed the custom Imperial triple combination pumper in service in February 1972, equipped with a 265 HP engine, Allison HT-740 automatic transmission, Waterous CMB-1 1,250-gpm single-stage pump on a 178-inch wheelbase. Standard features include air brakes with maxi-brake system and power steering. Optional features include stutter-tone air operated horns, electronic siren, 4-1/2 inch front suction, dual electric reels for 200-feet of 1-inch hose, 24-inch rear step and painted yellow lacquer. The rig was sold in February 1998 to Tower City, Pennsylvania - West End Fire Company Number 3 - Sheridan. *Photo courtesy of Lorne Ott*

Spec. No. I-537 Moorestown, NJ. - Lenola Fire Company received this first of its kind Model D-12 in June 1971. The chassis was manufactured by Imperial. Pierre Thibault Company built the body coachwork in Canada due to the fire department's rear discharge requirements. The rig, built on a 178-inch chassis, had a 6-71N, 265 HP engine, HT-740 transmission, a 1,250-gpm pump (CMBX) with a 6-inch front suction, and a 750-gallon tank. In 1987 Pierce Manufacturing of Appleton, Wisconsin, remanufactured the apparatus stretching out the wheelbase to 194 inches. The unit was sold in January 2001 to Jon's Truck Sales in Missouri. *Photo courtesy of Dave Ehrman*

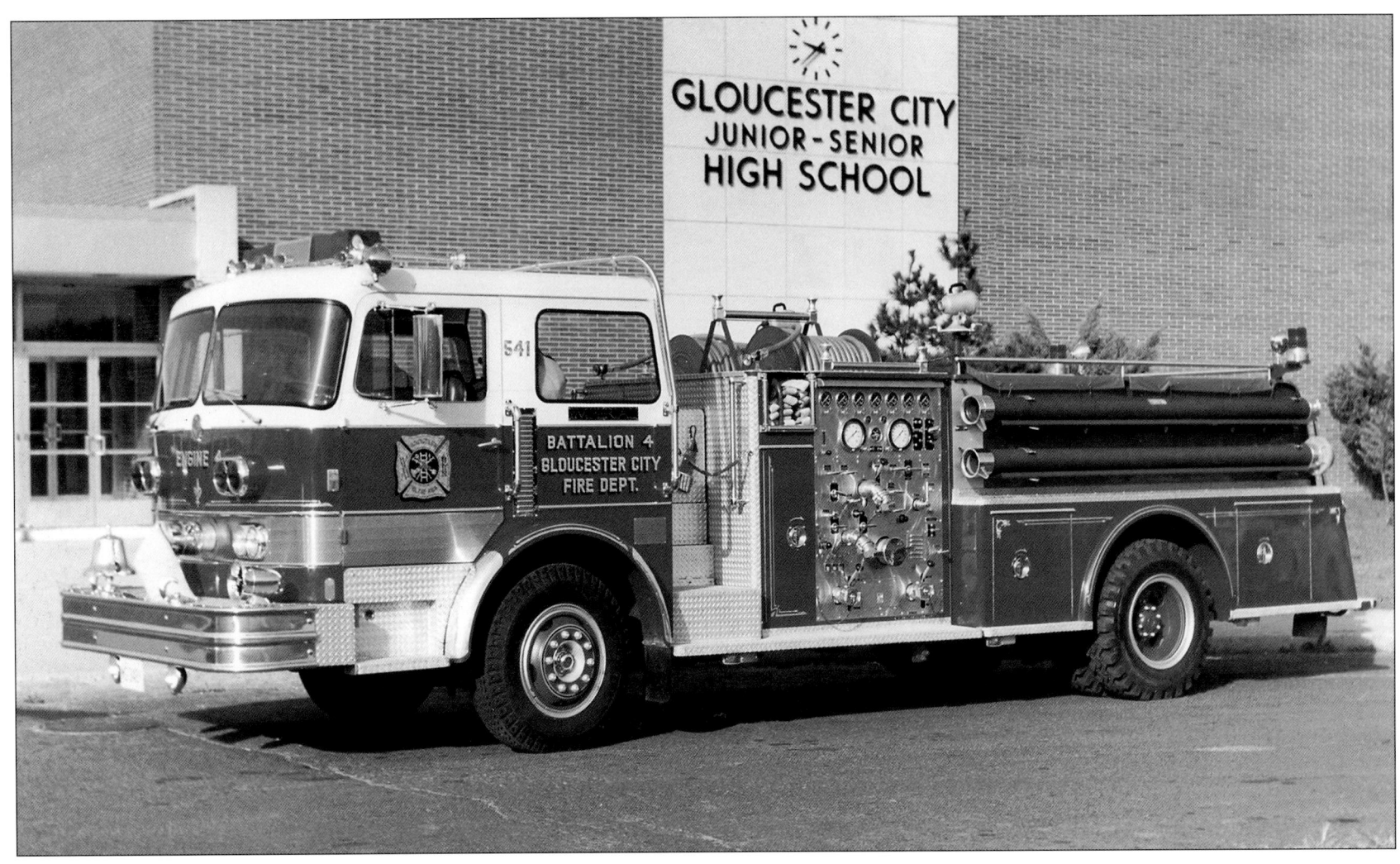

Spec. No. I-538 Gloucester City, NJ. - Gloucester City Fire Department took delivery of this cab-forward Model D-10 in July 1972. Its beauty was more than skin deep with "many value-added features," including a 12-inch fire bell, electronic siren, canopy cab assist rear rails (which would later be discontinued by the fire service), dual electric reel, large size compound and vacuum gauges, just to mention a few. A proposal price of $40,809 was presented on April 28, 1972. The unit served the city until August 8, 1992, when the Southern Stone County Fire Protection District in Reed Springs, Missouri acquired it—the first of four pre-owned Imperials in their fleet. It was traded in to Fire Master Inc. in April 2001 and sold to Rushville Fire Protection District, Missouri, on January 30, 2002. *Photo courtesy of Ralph Aspling*

Spec. No. I-539 Englewood Cliffs, NJ. - Englewood Cliffs Volunteer Fire Department placed their Model D-12 engine in service in June 1972. The 35,000-pound GVWR 178-inch pumper is powered with a Detroit Diesel 6-71N motor (265 HP at 2,300 rpm), a Spicer 6852K syncromesh manual transmission, and features a Waterous CMBX-1,250-gpm pump and a 500-gallon booster tank. Value-added features include a rear canopy handrail, dual electric recessed hose reels, NYC-style subway straps, left-hand wheelhouse compartments, plus many other miscellaneous items. After 30 years, the rig is still in service with the fire department today. *Photo courtesy of Ralph Aspling*

Spec. No. I-540 Sanatoga, PA. - Sanatoga Fire Company received their tanker in February 1972. The Ford C-900 tilt cab series, 175-inch wheelbase has a Ford 534 V-8 (276 HP) gas engine and a Spicer 6352 transmission. The 38,000-pound GVWR single rear axle chassis carries 2,000-gallons of water and has a Waterous CPK-3, 500-gpm pump and is equipped to set up a Porto-pond or dump into one. It is equipped with 500 feet of 2 1/2-inch supply hose and has one 1 1/2-inch hose lay in the bed. The contract price was $15,295 as proposed on April 3, 1971. Refurbished in 1994 by a local body shop, the tanker is still in service with the department today. *Photo courtesy of Don Woodly*

Spec. No. I-541 Seaside Heights, NJ. - Seaside Heights Fire Department took delivery of the Model D-12 in May 1972. The 35,000-pound GVWR chassis is powered by a Detroit Diesel 6-71N motor, with an Allison automatic transmission. Many value-added features include a two-stage Waterous 1,250-gpm pump, a 6-inch front suction, extended front bumper with soft suction hose compartment, and a cross-mounted electric reel. The unit was sold to Fire Department "Bomberos Piedras Negras, Coah" Mexico, in May 1996, "after 24 years of good service," stated current Chief Joe Paolo. *Photo courtesy of Ralph Aspling*

Spec. No. I-542 Windsor, CT. - The Poquonck Fire Department received this Imperial Model AS-7 Mack Model MB-685S Cab Over Engine special pumper tanker in October 1972 at a bid price of $47,847. The 46,000-pound GVWR chassis built in March 1972 is equipped with a Maxidyne diesel engine, 206 HP, ENDT 675, and a Maxi-torque TRL 107 transmission, with 12,000-pound front axle and tandem 34,000-pound rear axle. The pumping unit was an engine-driven (Chrysler V-8, 180 HP) Hale 45FA-C318, 750-gpm and was supported by a 2,000-gallon sloping bottom tank, plus many other valued features. The rig serves the department today. *Photo courtesy of Mark Redman*

Spec. No. I-543 Allentown, PA. - Western Salisbury Fire Company took delivery of this mini pumper in May 1972. The fire-fighting package was installed on a Ford F-350 series cab and chassis and included a Waterous power take-off pump, a CPK-3 500-gpm pump and a 250-gallon booster tank. A cross lay hose compartment was provided at the front of the body. Telescoping floodlights were body mounted. The unit was sold in the spring of 1988 to the Sandts Eddy Fire Company in Pennsylvania, which later merged with Lower Mount Bethel Fire Department in Martin's Creek, Pennsylvania. Eventually a booster reel was added. *Photo courtesy of Todd Lincoln*

Spec. No. I-544 Barrington, NJ. - The Barrington Fire Company Number 1 placed the Model D-12 in service in September 1972. The triple combination pumper is powered with a Detroit Diesel 8V-71N engine, Spicer transmission, 12,000-pound front and 23,000-pound rear axle. A Waterous 1,250-gpm pump and a 500-gallon water tank are installed. The rig has many optional items including a Federal Q2B electronic siren, dual cab spotlights, Federal bar lamp, air horns, and high driver side compartments. The engine served Barrington until November 1993 when it was sold to the Cherry Creek Volunteer Fire Department, Sparta, Tennessee, where it serves the community today. *Photo courtesy of Dave Ehrman*

Spec. No. I-545 Wappingers Falls, NY. - The New Hamburg Volunteer Fire Department took delivery of their Model D-12 in February 1972. The custom cab-forward triple combination pumper was built on a 178-inch wheelbase, and powered with a Detroit Diesel engine (265 HP), it uses a 5-speed Spicer syncromesh transmission. The midship-mounted pump is a Waterous two-stage 1,250-gpm pump and the booster tank capacity is 500 gallons. The all-yellow unit is equipped with a 6-inch front suction and also features top-mounted air horns and an electronic siren. The rig sold to the Nelsonville Volunteer Fire Department in New York on July 30, 1997. *Photo courtesy of John Perusse*

Spec. No. I-546 Bayville, NJ. - Bayville Volunteer Fire Department placed their custom pumper tanker Model D-10 (Waterous CMBX) in service early in 1972. The 8V-71 chassis and body was refurbished by Hahn Fire Apparatus (which ceased operation in December 1989), in 1984. The 245-inch wheelbase, 2,000-gallon tanker was refinished from its original fire department red color to lime-yellow and dark green. Northeast Fire Apparatus, Inc. first sold this unit in May 1994 to the North Oldham Fire Department of Goshen, Kentucky. It was sold again in September 1996 to the Burkesville Kentucky Fire Department. *Photo courtesy of Weldon Rowe*

Spec. No. I-547 Toms River, NJ. - Manitou Park Fire Company, Berkley Township received their Model D-10 in April 1972. The fire department red and white 1,000-gpm Waterous (CMBX) custom pumper is supported with a 1,000-gallon water tank. It features a Detroit Diesel 265 HP engine with a Spicer 6852 transmission on a 178-inch wheelbase. Optional extras include triple boot and coat rails on each side, fire bell, electronic siren, cab spotlights, three 10-foot lengths of 5-inch hard suction hose, a portable deluge gun and two portable floodlights. The rig has served the community for over 30 years and is with the department today. *Photo courtesy of Ralph Aspling*

Spec. No. I-548 Barnegat, NJ. - Barnegat Volunteer Fire Department accepted their Imperial Model D-10 in April 1972. Detroit Diesel re-rated the 6-71N to 325 HP, matched with the Spicer 6852 transmission and a Waterous two-stage CM 1,000-gpm pump with relief valve. The pump transfer valve was electric with manual overrides. A contract price of $40,975 was signed on June 30, 1971. Special value-added features included rear shocks, 5-inch front suction with air operated panel-controlled butterfly valve, triple coat and boot racks on both sides of the hose bed, and a 1,000-gallon water tank. The unit remains with the department today. *Photo courtesy of Ralph Aspling.*

Spec. No. I-549 (BTB) - Montreal, Quebec, Canada. - The Montreal Est. Fire Department received the first Imperial chassis to be sold to Pierreville Firetrucks that delivered the 100-foot Thibault rear-mounted aerial, in June 1972. The 6-71N-powered unit has a Spicer 6852 syncromesh transmission. Confirming the chassis number has not been successful but it is (BTB) "believed to be" the number identified. The original truck was painted fire department red and repainted in later years to sunflower yellow. It is currently in service. Unfortunately all data plates have been removed. *Photo courtesy of Ken Walton*

Spec. No. I-550 Pierreville, Quebec, Canada - Gloucester Fire Department, Ontario, Canada received this Pierreville (PTF-270) aerial unit built on an Imperial 220-inch wheelbase chassis in October 1972. The 40,000-pound GVWR, 16,000-pound front axle, 24,000-pound rear axle, 6V-71, 240 HP aerial truck has a Thibault midship-mounted 100-foot 4-section hydraulic ladder, Allison HT-70 transmission and a Waterous 750-gpm (840 Imperial Gallons) single-stage pump. No booster tank was installed. In August 1989 the unit was sold to the Kapuskasing Fire Department where it remains in service. *Photo courtesy of Ken Walton*

Spec. No. I-551 Jenkintown, PA. - Pioneer Fire Company placed their Model D-10 in service in April 1972. The rig's motor was a gas powered Waukesha F817G, matched with a Spicer 6852 manual syncromesh transmission and a 1,000-gpm Waterous two-stage pump. Features included a 500-gallon water tank, dual 1-inch recess-mounted electric reels (one in each rear body side compartment), a 30-foot extension ladder, a 14-foot roof and 10 folding ladders, three extinguishers, two 10-foot lengths of hard suction hose and a fire bell, a Federal Q siren, rear canopy rail and other items. The unit was sold to Combine, Texas in March 1996. *Photo courtesy of Gary Bachman*

Spec. No. I-554 Brewster, NY. - Brewster Fire Department received this Imperial Model B brush truck on a Ford F-150 chassis on March 19, 1972. The booster pump is a Waterous CPK-1, 60-gpm, and it has a booster tank of 200 gallons. Value-added features include dual electric hose reels, extended front bumper with electric mounted winch, a portable pump, and storage for miscellaneous brush fire equipment (i.e., brooms, portable Indian tanks, and assorted brush tools, including rakes and shovels). The quick response unit is still in active service and has served the community for over 30 years. *Photo courtesy of Ron Bogardus*

Spec. No. I-555A Burlington Township, NJ. - Independent Fire Company Number 1 placed this Model D-10, one of three units purchased for the three township's fire departments (Beverly Road, Relief Fire Co.) in service on June 30, 1972. Each unit featured the same apparatus and was painted a different color. Each featured a Detroit Diesel 8V-71, 350 HP at 2,300-rpm engines, manual Spicer 6852 transmissions, 1,000-gpm Waterous CMBX pumps and 1,000-gallon tanks. Purchased by a truck broker, Jon's Equipment Sales in Rogersville, Missouri, it was resold to Southern Stone County Fire Department, Reed Springs. Missouri on February 9, 1994, the third pre-owned Imperial purchased. *Photo courtesy of Jerry Bell*

Spec. No. I-557 Plattsburg, NY. - South Plattsburg Fire Department received their commercial Ford C-900 tanker in September 1972. The short 161-inch, 48,000-pound GVWR tandem chassis, 12,000-pound front and 36,000-pound rear axle, is gas powered with a Ford 534 (266 HP) engine and a manual transmission. The pump is a Waterous single-stage 1,000-gpm and has a 2,000-gallon water tank. Three 10-foot lengths of hard suction hose were supplied. Note the doublewide compartment aft of the rear wheels. The yellow unit remained in the department until October 1977 when it was sold to the Duane Volunteer Fire Department in Duane, New York. *Photo courtesy of Dave Ehrman*

Spec. No. I-558 Forest City, PA. - Forest City Volunteer Fire Department took delivery of the Model A-10 in September 1973. The Ford Model C-900 featured a 175-inch wheelbase, 148-inch CA, 34,000-pound GVWR, with a gas 534-SD V-8 engine, Allison MT-42 automatic transmission, 12,000-pound front axle and a 23,000-pound rear axle. The pump size was a CMBX 1,000-gpm two-stage and the tank was 1,000 gallons. Amthor's of Walden, New York added the body high side compartments in 1900, before the unit was sold to Grand Bay, Alabama in 1986, and again to McCollum Midway Fire Department in Jasper, Alabama in 1998. It is in service today. *Photo courtesy of Paul Lukas*

Spec. No. I-559 Mattituck, NY. - Mattituck Fire District received their Imperial Model AS-7 Ford C-950 in October 1972. The gas powered 534-SD V-8 engine was matched with a Waterous CMBX 1,750-gpm single-stage pump. It had a wheelbase of 175 inches, a GVWR rating of 34,000-pounds with a 12,000-pound front and a 22,000-pound rear axle, air brakes, dual battery system, 1,000-gallon booster tank, one 10-foot length of hard suction hose, a 30-kw generator with 110 and 220 outlets and 500-watt floodlights, Morton Kass steel running board and rear step. The apparatus was sold to a broker, North East Fire Company, in January 1995, and then to Marquand, Missouri. *Photo courtesy of John Keogh*

Spec. No. I-562 North Banford, CT. - North Banford Fire Company took delivery of their International Harvester "Model 1310," four-wheel drive mini pumper in mid 1972. The pump and roll unit was equipped with two booster reels with 300 feet of 3/4-inch nozzles, hard suction hose, Kussmaul battery charger, Milo engine brake, 250-gpm pump, 250-gallon water tank, NFPA reflective striping, siren, warning light and light bar. The rig was first sold to Pine Rock Park Fire Company in Shelton, Connecticut and later on August 28, 1999, to Elderado, Ohio. *Photo courtesy of Firetec*

Spec. No. I-566 Woodbury, NJ. - Goodwill Fire Company Number 2 accepted their Model D-12 on December 22, 1972. The 5-man canopy cab, 39,000-pound GVWR Imperial chassis had a Detroit Diesel 8V-71 with a Spicer manual transmission, 16,000-pound front axle and 23,000-pound rear axle. Pump size was 1,250-gpm and the tank size was the standard 500-gallons. It also featured pre-connected front suction, right-hand chrome plated bell, mechanical siren, portable deluge gun, electric reels, and triple boot and coat rails. The original cost presented on December 7, 1971, was $50,903. The unit was refinished by Grumman Emergency Products and is still with the department today. *Photo courtesy of Skip Stinger*

Spec. No. I-568B Warminister, PA. - Warminister Fire Department purchased two Imperial engines, which were delivered in January 1972. The Model D-12 custom rigs are powered with Detroit Diesel 6-71N engines and Spicer transmissions, Waterous 1,250-gpm CMBX-1 pumps and 500-gallon booster tanks. Value-added features included additional warning and lighting equipment on the front (i.e., a fire bell, Q2B electronic siren, oscillating white light, and large red warning lights). Unit I-568B was sold in October 1985 to Canton, Pennsylvania and served that community until July 2001 when it was placed in reserve. Unit I-568A was sold to Roulette, Pennsylvania. *Photo courtesy of Todd Lincoln*

Spec. No. I-569 Ridge, NY. - The Ridge Fire District took delivery of the Model D-10 Imperial triple-combination pumper on April 18, 1974. The 196-inch wheelbase (because of the 36-inch locker) was powered by a 6-71N Detroit, an Allison HT-740 transmission, and had a GVWR rating of 35,000 pounds. A Waterous CMBX 1,000-gpm and a CPK-4 (60-gpm) hi–pressure pump, and a 1,000-gallon tank were provided. Body compartment-mounted electric reels were provided. The rig was sold on December 23, 1994, to Prattsville Fire District, New York. In October 2000 the engine failed. It was then sold for its usable components to Ashland Sand & Gravel Company in Ashland, New York. *Photo courtesy of Ralph Aspling*

Spec. No. I-570 Shavertown, PA. - The Shavertown Fire Department received their Ford C-900 Imperial pumper in November 1972. The 153-inch wheelbase, 3-man tilt cab-over is powered by a Ford 534, 277 HP gasoline engine with a manual transmission on a 34,000-pound GVWR chassis. The pump is a single-stage Waterous CMBX 1,750-gpm and the tank has a 1,000-gallon capacity. Extras include warning lights, siren, cross lays, NFPA equipment, automatic suction valve, hose bed cover, and many other extras. The piece was sold to Courtdale, PA in 1987, then to Haneyville in 1995. On May 4, 1999, it was sold to Jasper, Alabama for $9,000. *Photo courtesy of Firetec*

Spec. No. I-571 Hauppauge, NY. - Hauppauge Fire Department placed their Model D-12 in service in January 1973. A Cummins diesel NTF-295 powered the 35,000-pound GVWR rig. It featured a 196-inch wheelbase for a double-wide (36 inch) transverse compartment, an Allison HT-740 transmission, and a Waterous CMBX 1,250-gpm two-stage pump with a tank capacity of 500 gallons. The all-white finished apparatus was sold to the North Hopewell/Winterstown Fire Department of Felton, New York on October 14, 1991. In June 1992, a red lower half was added. It is in service today as a front-line piece, and mostly runs on auto accidents and as a secondary engine on structure fires. *Photo courtesy of Warren Gleitsmann*

Spec. No. I-574 Westmount, NJ. - Fire District Number 1, Haddon Township, purchased this Imperial Rescue unit in February 1972 on an International Harvester Cargo Star, Model CO-1910A. It had a 169-inch wheelbase chassis, a FTV-549 gas engine, and a T-112 automatic transmission. It included a 15-foot rear walk-in rescue at a cost of $42,375. Delivery was made early in 1973. The 2-man cab was extended with a canopy that allowed for four rear-facing crew seats. Loose equipment included an Onan Model 6.5 electric generator, cable reel, and many value-added features. The rig was sold to Deerfield Volunteer Fire Department in Virginia and was later resold to a private party. *Photo courtesy of Dave Ehrman*

Spec. No. I-576 Lauderhill, FL. - Lauderhill Fire Department placed their Model D-4-S, the first Imperial rear-mounted Thibault 4-section 100-foot stick, in service in June 1973. It featured the low profile 230-inch cab chassis wheelbase, a 239 HP Detroit Diesel 6-71N motor and a Spicer syncromesh transmission. Beauford, South Carolina purchased the apparatus in July 1983, returned the rig to Pierreville Fire Trucks in Quebec, refurbished the unit to 1983 standards and added a pre-piped 3-inch dual telescoping waterway, a 1,000-gpm electric nozzle, upgraded the hydraulic system, and repainted the unit. It was placed in service in February 1984. *Photo courtesy of Jon Umbdenstock*

Spec. No. I-577 Kissimmee, FL. - Kissimmee Fire Department took delivery of their Model D-12 engine in December 1973. The request for quotation submitted on January 17, 1972, listed a contract price of $40,347. The 178-inch wheelbase, 35,000-pound GVWR apparatus, powered with an 8V-71 350 HP engine, Spicer 5-speed transmission had a Waterous CSUMBX with pumping capacity of 1,250-gpm. The tank size was 750 gallons. Lighting equipment included a recessed Mars 888 light. The rig was originally painted yellow. It was eventually changed to red and white. Due to its age, this rig was relegated to reserve status but still serves the department today. *Photo courtesy of Mike O'Dowd*

Spec. No. I-579 Cornish, NH. - The Cornish Fire Department placed their Imperial pumper on a Ford C-900 chassis in February 1973. The 175-inch wheelbase is powered with a Ford 534 gas engine (266 HP) and a manual Spicer 6352 transmission. The midship pump is a Waterous 750-gpm Model CMBX-750 with a 750-gallon capacity tank. The triple combination pumper served the community until March 1995 when Dingee Machine Company, a regional fire apparatus manufacturer, delivered a new apparatus and placed the Imperial in the company's inventory. *Photo courtesy of Bob Rice*

Spec. No. I-580 Allentown, PA. - Greenawalds Fire Company, South Whitehall Township, accepted their Ford Imperial pumper in service in 1972. The 3-man tilt cab has a wheelbase of 175-inches, a Ford 534 (266 HP at 3,200 rpm) motor and uses a 1,000-gpm Waterous CMBX pump and a 500-gallon tank. Note discharge fittings are angled down for ease of hose connection. The bumper has been extended to accommodate 5-inch soft suction hose. Five air packs were hung on the left-hand side, and the hard suction hose was moved to each side, above the hose body. The unit was sold in 1995 to a broker in Alabama. *Photo courtesy of Todd Lincoln*

Spec. No. I-582 Ship Bottom, NJ. - Ship Bottom Volunteer Fire Company Number 1 Inc. received this unique four-door cab on a Ford C Model/Series 3-man tilt cab 27,000-pound GVWR chassis in November 1972. The motor is a Ford 534, with a manual Spicer transmission, a 9,000-pound front axle and an 18,500-pound rear axle. The rear stationary cabin has two 64-inch-long seats (one rear and one forward facing), a 750-gpm pump and a 500-gallon booster tank. Over the wheelhouse body compartments are a front suction and a rear-mounted hose reel. In December 2000 the unit found a good home when it was donated to the Chatsworth Volunteer Fire Company in New Jersey. *Photo courtesy of Scott Mattson*

Spec. No. I-583 Magnolia, DE. - Magnolia Fire Volunteer Company took delivery of their Model D-10 in 1973. The sunflower yellow finished apparatus is powered with a Detroit Diesel 6-71N, 238 HP engine, and an Eaton 5-speed transmission on a 180-inch wheelbase. The front axle holds 12,000 pounds and the rear axle is a 2-speed electric shift Eaton with 23,500-pound capacity. It features a Waterous 1,000-gpm pump with an extended tank and body compartmentation. Note a double door compartment is provided in front of the rear wheels. The engine is in service, but on reserve status. *Photo courtesy of Dave Ehrman*

Spec. No. I-584A-B-C-D-E Fairfax, VA. - The Fairfax County Fire Department purchased five Imperial pumpers in 1972, which were delivered in 1972 and 1973. This was to be one of the largest orders for apparatus received by Imperial. The 35,000-pound GVWR engines are equipped with Detroit Diesel 8V-71N engines and Spicer syncromesh Model 6853B transmissions. The units have 1,000-gpm pumps with front suctions and 500-gallon water tanks. Emergency One rehabbed all units. No information is currently available as to their whereabouts. *Photo courtesy of Ralph Aspling*

Spec. No. I-586 Medford, NJ. - The Lakes Fire Company in Medford Township received this unique Model D-10 pumper in May 1973. The fire department changed its name to the Taunton Fire Company in 1985. The 46,000-pound GVWR is powered with a 220 HP 6-71N Detroit Diesel engine, a 5-speed Spicer 6853B syncromesh transmission, and a tandem axle with wide single tires enabling the rig to go where regular rigs couldn't go. The midship pump is a Waterous 1,000-gpm two-stage Model CMBX with a 1,000-gallon tank. In December 1989 the Hidden Valley Volunteer Fire Department in Sparta, Tennessee purchased the rig. *Photo courtesy of Ralph Aspling*

Spec. No. I-587 Wayne, PA. - Radnor Fire Company took delivery of their Model D-10 in February 1973. The 35,000-pound GVWR, 178-inch rig is powered with a 6-71N engine (265 HP), Allison HT-70, 5-speed automatic transmission and has a 12,000-pound front axle and a 23,000-pound rear axle. It features a single-stage Waterous CSK 1,000-gpm pumping capacity with a 500-gallon water tank. Six pre-connected hose lines are provided and the hose bed carries 1,000 feet of 5-inch hose. The original color of the engine was a white cab roof with a yellow lower half. The apparatus was sold to the Maytown East Donegal Township Fire Department in Pennsylvania in May 1984. *Photo courtesy of Ralph Aspling*

Spec. No. I-587 Maytown, PA. - A before and after photo of Maytown East Donegal Township apparatus after it was re-cabed and refurbished by Tri-State in Trenton, New Jersey in December 1989. The color was repainted to white over red. Pemfab Trucks in Rancocas, New Jersey installed a four-door cab and extended the wheelbase to 184-inches after the apparatus was involved in a head-on collision with a drunk driver in 1988. The pump and entire rear remained the same; a 500-gallon poly tank replaced the metal tank. The engine is sent on approximately 300 calls per year, and is scheduled for replacement at the end of 2002. *Photo courtesy of Dave Ehrman*

Spec. No. I-588 Forest City, FL. - The Forest City, Bear Lake Fire Control District, took delivery of the Model D-12 on October 27, 1972. The 35,000-pound GVWR has a 318 HP motor, syncromesh transmission, Waterous 1,250-gpm CMBX midship pump and a 750-gallon water tank. Seminole County EMS/Fire/Rescue Department took control of the fire district and was consolidated in October 1974. The unit saw front-line duty until 1990. It was placed in reserve status and is seldom used today except for special occasions. The original rig was finished in the traditional red paint but now is lime-yellow and white. *Photo courtesy of Ralph Aspling*

Spec. No. I-589 Havertown, PA. - Oakmont Fire Company Number 1, Haverford Township, received their Model D-12 in June 1973. Built on the standard 178-inch wheelbase, the 5-man canopy cab pumper is powered with a Detroit Diesel 6-71N in-line engine, and coupled with a Spicer syncromesh Model 6852 transmission with a Spicer 14-inch dual plate clutch. The front axle is a Timken/Rockwell FF-901 (12,000 pounds), and the rear axle is a Timken/Rockwell R-140 (23,000 pounds). A 500-gallon tank supported the Waterous 1,250-gpm Model CMBX pump. The unit was sold in 1983 to Campbelltown, Pennsylvania. *Photo courtesy of Dave Bowen*

Spec. No. I-589 Campbelltown, PA. - After the refurbishment of this rig in 1999 by the Pennsylvania Fire Apparatus Company in Gettysburg, it deserves a second look. The canopy cab was modified to a 7-man closed cab by adding seating capacity for two additional fire personnel, and two additional rear cab doors. Other modifications included square headlamps and warning bezels, recessed siren, recessed air horns and speakers. The apparatus was completely refinished and white reflective striping was added. All of these modifications were done on the original 178-inch wheelbase. After 30 years, the engine is second due on all structure fires and first due on all accidents with injury responses. *Photo courtesy of Michael J. Breive*

Spec. No. I-590 Asbury, NJ. - Asbury Fire Company took delivery of their Model D-10 in June 1973. The 36,000-pound GVWR Model D-10 is equipped with a 12,000-pound front axle and a 24,000-pound rear axle, a 265 HP diesel engine, syncromesh transmission, 10.00 x 20 tires, and a 1,000-gpm Waterous CSMV single-stage pump. Other features include a 1,000-gallon tank, pump and roll capability through a secondary 250-gpm PTO pump, portable deck gun, 2,100-gallon Porto tank, electric hose reel, two 500-watt floodlights and NFPA equipment. The contract price on June 21, 1972, was for $39,900. It was refurbished in 1991 and sold in March 2000 to Nash Fire Department, Oklahoma. *Photo courtesy of Gary Wood*

Spec. No. I-591 Millville, NJ. - The Millville Fire Department received their Model A-4S Ladder truck in August 1973. The 48,000-pound GVWR rig has a Grove 4-section 100-foot rear-mounted aerial ladder, 8V-71N engine, Allison HT-740 automatic transmission, 14,000-pound front axle and a tandem 34,000-pound rear axle, with a low profile 5-man canopy closed cab. In 1990 the rig was sent to Ladder Towers in Ephrata, Pennsylvania where it was completely rehabbed. The original finish was yellow with a gray ladder. In 1994 the unit was repainted red with a white upper cab and a white ladder. After 30 years, the unit remains in service with the department. *Photo courtesy of Steelman Hines*

Spec. No. I-596 Lawnside, NJ. - Lawnside Fire Company took delivery of this International Harvester tilt cab pumper in September 1974. The 35,000-pound GVWR, 168-inch wheelbase Cargo Star commercial chassis, Series CO 1910B, was powered with an FTV-549 gas engine, and a Spicer transmission. The IHC tilt cab provided seating for three firemen. A 73-inch-wide canopy cab was added by Imperial and provided additional seating for three more firemen. A Waterous 1,000-gpm two-stage pump was supplied along with a 750-gallon water tank. The engine failed in December 1994, and the rig has been stored at Dom's IHC dealer in Cherry Hill, New Jersey ever since. *Photo courtesy of Ralph Aspling*

Spec. No. I-597 Glenham, NY. - The Stlater Chemical Fire Company accepted their brush truck in April1972 on a 3/4-ton Chevrolet four-wheel drive chassis. It has a Waterous CPK-1 (100-gpm at 200 psi) PTO pump, with a 200-gallon water tank and 200 feet of 1-inch booster hose on an electric hose reel. The brush rig was sold to the Plattekill Fire Department in September1989. Chief Brugger stated, "The vehicle has seen a lot of action over the years for us including numerous mountain fires and is the first out for all storm-related calls." *Photo courtesy of Ron Boardgus*

Spec. No. I-598 Catasauqua, PA. - The Catasauqua Fire Department placed their Model D-12 in service on January 14, 1974. The all-yellow 36,000-pound apparatus was equipped with a 6-71N engine, 6852 transmission, a Waterous CMBX 1,250-gpm pump and a 500-gallon tank. The purchase price was $54,265. Optional extras included full height wheelhouse compartments on both sides, a compartment under the hose bed for two 10-foot hard suctions, a 14-foot roof, 35-foot and 50-foot extension ladders, a 10-foot folding ladder, and a 6-inch front suction. Micro Fire Apparatus in Allentown, Pennsylvania completed the rehabilitated unit in April 1992. *Photo courtesy of Samuel Burrows*

Spec. No. I-599 Mendham, NJ. - The Independent Hook & Ladder Company acquired this Imperial Quad that was delivered in October 1973. The 196-inch chassis is powered with a Detroit Diesel 8V-71N, 350 HP engine, Allison HD-740 transmission, and has a Waterous CMBX 1,000-gpm pump, a 500-gallon water tank and 184-feet of ground ladders. Additional features include a 20-gallon foam tank, high-pressure capacity, 1,500-feet of 3-inch hose, and two 300-foot sets of 1-1/2 inch lines. A broker, Emergency Equipment Service of Augusta, Georgia, sold the apparatus in January 1992 to the Leah Fire Department in Appling, Georgia. *Photo courtesy of Ken Betz*

Spec. No. I-600 Orangeville, PA. - The Orangeville Community Fire Company received their Model D-7 in July 1973. Thc apparatus was powered by a Waukesha F-817G, 318 HP gas engine matched with a Spicer Model 6852 syncromesh transmission and had a Waterous CMB-750, a CPK-3 150-gpm booster pump and a 750-gallon booster tank, plus a 150-gallon foam storage tank. A large size fold-a-tank was mounted on the left-hand side. This apparatus served the community for nearly 23 years. In February 1996 ownership was transferred to Brewer Equipment Company for salvage. *Photo courtesy of Ralph Aspling*

Spec. No. I-602 Auburn, ME. - Auburn Fire Department took delivery of their Model D-10 triple combination custom pumper on July 23, 1973, and within two hours responded to their first fire call. The diesel-powered 265 HP, 6-cylinder 6-71N was matched to an HT-740 automatic transmission on a 184-inch wheelbase. Pumping capacity was a two-stage Waterous CMBX 1,000-gpm pump and a 500-gallon booster tank. Optional extra features included roof-mounted air horns, Federal Q mechanical siren, and dual top-mounted hose reels. The unit is still in service today at Auburn's Volunteer Danville Fire Station. *Photo courtesy of Ralph Aspling*

Spec. No. I-603 Pennsauken, NJ. - Bloomfield Park Station Number 3 placed their Model D-10 pumper in service in September 1973. The 35,000-pound GVWR unit was powered with a Detroit Diesel 8V-71N, a Spicer 5-speed Model 6852 syncromesh transmission, with a Waterous 1,000-gpm pump and a 500-gallon tank. The unit was in service 20 years before Pierce Manufacturing Company of Appleton, Wisconsin refurbished it in April 1993. Original valued extras included a front suction, fire bell, Federal Q-2B siren, dual beacon ray lights, dual-mounted air horns, hose reel, cross lays, and portable hand lights. *Photo courtesy of Dave Ehrman*

Spec. No. I-605 Saint James, NY. - Saint James Fire Department received their Model D-12 Imperial 5-man (84-inch-wide cab) custom triple combination pumper in September 1973. The 178-inch wheelbase rig had Cummins power, a NHF-265 diesel, an Allison automatic transmission MT-644, a Waterous CMBX 1,250-gpm two-stage pump, a CPK-4 (60-gpm at 1,000 psi) 1,000-gallon water tank, three wheelhouse compartments, Duo-Safety ground ladders, and was finished red and white. Ownership was transferred to a broker, Northeast Fire Apparatus Inc., and it was sold to Belvue, Kansas in March 1993. *Photo courtesy of Northeast Fire Apparatus Inc.*

Spec. No. I-608 Shirley, MA. - The Commonwealth of Massachusetts Industrial School for Boys purchased this model D-10 for the fire department. Bid price was $40,000. The triple combination pumper has a Waterous CM 1,000-gpm two-stage pump, a 600-gallon tank, a no-spin 24,000-pound rear axle, 1,200-watt DC transformer, air operated radiator shutters, Perry water filter, in-line fuel shutoff, individual pressure gauges for the six 2 1/2-inch and two 1 1/2-inch outlets, speaker grille and mike compartment on pump panel, and includes two 1 1/2-inch Mattydale cross lays over the operating panel. The apparatus was refurbished in 1991 by the E. J. Murphy Company and is in service today. *Photo courtesy of Richard J. Bartlett*

Spec. No. I-609 Owings Mills, MD. - Owings Mills Volunteer Fire Company received their Model D-4-S aerial unit in July 1974. The low profile, 84-inch-wide Cincinnati cab, 52,000-pound GVWR chassis front axle is 16,000 pounds and the rear tandem is 36,000 pounds. The engine is a Detroit Diesel 8V-71, with an Allison HT-750 transmission, and the aerial ladder is a 4-section LTI. The cab was redone in 1986 by 3D Manufacturing (purchased by American LaFrance in April 1999). Duplex re-chassised the unit in 1996 and added a four-door cab, utilizing the major components (engine, transmission and axles). It is in service today at Owings Mills. *Photo courtesy of Mike Sanders*

Spec. No. I-611 Grenloch, NJ. - The Washington Township Fire Department received their Model A-4-S in August 1974 at an original cost of $119,712. The GVWR of 60,400 pounds required an 18,000-pound front and a tandem rear axle for 42,000 pounds using an 8V-71N engine and an Allison HT-740 automatic transmission. A PTO Waterous CPK-3 (400-gpm) two-stage centrifugal was provided on the LTI 4-section 100-foot rear-mounted ladder. The unit contained 280-feet of ground ladders. In November 1992 the two-tone red and white apparatus was sold to the Friendship Fire Company of Mohnton, Pennsylvania. *Photo courtesy of Dave Ehrman*

Spec. No. I-612 California, MD. - Bay District Volunteer Fire Department (Formerly Lexington Park Volunteer Fire Department) received their Model DA-4S in February 1975, at a cost of $111,036. The 54,000-pound GVWR chassis on a 230-inch wheelbase with a 16,000-pound front axle and a 38,000-pound tandem rear axle is powered with an 8V-71 Detroit Diesel engine and an Allison automatic transmission. The ladder is a Ladder Towers 100-foot 4-section rear-mounted ladder. Generous body compartmentation is provided. American LaFrance, Pemfab Trucks and Ladder Towers rehabbed the complete unit from September 1990 through May of 1992. *Photo courtesy of Mike Sanders*

Spec. No. I-613 Windsor, VT. - The Windsor Fire Department placed their Model D-12 in service in December 1973. The bid in January 1972 was $33,554. Provided on the wheelbase of 178 inches was a Detroit Diesel 6-71N, 265 HP engine and the standard Spicer 6852 transmission. The pump was a Waterous 1,250-gpm CMBX model with a 750-gallon water tank. The unusual "suicide door" on the front left-hand body compartment provided unobstructed access to tools mounted in the front compartment. Optional extras included an overhead ladder mounting, a portable deluge gun, and three hard suction sleeves, to name a few. The rig remains in service today. *Photo courtesy of Ron Bogardus*

Spec. No. I-614 Rochelle Park, NJ. - Rochelle Park Volunteer Fire Department took delivery of their Model D-12 in September 1973. The low profile 5-man canopy cab was required due to a height restriction on the firehouse doors. The lime-yellow colored apparatus was powered by a 6-71N motor, a syncromesh transmission and had a pumping capacity of 1,250-gpm with a Waterous CMBX pump. Tank size was 500 gallons. The unit was moved into a new firehouse completed in 1978 and a deck gun and platform were added along with the raising of the booster reels. The rig was sold in June 1999 to the Bergen County Fire Academy in Mahwah, New Jersey. *Photo courtesy of Michael Martinelli*

Spec. No. I-616 Levitown, PA. - Bristol Townships Edgely Fire Company received their Model D-12 in October 1973. The two-tone red lower half and the white canopy cab unit was powered with a Detroit Diesel 8V-71N, 350 HP engine and matched with a Spicer Model 6852 transmission. The Waterous CMBX 1,250-gpm pump was outfitted with angled discharge adapters for ease of hose line hook-up. The water tank capacity was 500 gallons. Numerous value-plus options were provided and in 1989 the custom rig was sold to Union Volunteer Fire Department Number 1 in Morrisville, Pennsylvania. *Photo courtesy of Michael Martinelli*

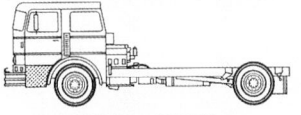
Little advertising existed in trade publications until Imperial's last advertisement appeared in September 1973 when the company, having made the decision, announced "Specify Imperial Chassis." The ad featured a custom Imperial fire apparatus foe. The apparatus for Radnor, Pennsylvania (I-587) was delivered in February 1973, and for Medford Township, New Jersey (I-586) in May 1973. While orders for sales in process were honored and manufactured, it was clear to all in the industry that the company's direction had changed.

Spec. No. I-617 Manchester Township, NJ. - The Ridgeway Volunteer Fire Company purchased this Pierce "Fire Marshall" Spec. No. 8324-C apparatus built on an Imperial custom chassis. This sale marked the forward decision to market "chassis only" by the company. The chassis manufactured in June 1974 was delivered to Appleton, Wisconsin for completion. The power is an 8V-71N engine and a Spicer 6852 transmission. Pump size is 1,250-gpm and the tank size is 500 gallons. The yellow rig served the department until July 2000, when the Lebanon Lakes Fire Department in Chatsworth, New Jersey, purchased it. They then repainted the unit red and white. *Photo courtesy of Scott Mattson*

Spec. No. I-619 Vancouver, British Columbia, Canada - The Vancouver Fire Department received one of the four engines purchased from Pierreville (PTF-456) on Imperial Fire Chassis on August 20, 1974. This engine (A-9138) had a Detroit Diesel 265 HP, 6V-71 motor, an Allison transmission, a Waterous 1050 Imperial gpm pump, a 300-gallon water tank and a 20-gallon foam tank, and was sold at auction on October 14, 1999. The unit A-9156 (PTF-455), in service October 24, 1975, was sold October 14, 1999; A-160 (PTF-605) was sold in 1996, and A-184 (PTF-455), in service on August 14, 1974, was sold in 1994. Little is known about where they are today. *Photo courtesy of Shane Mackichan*

Spec. No. I-621 Vancouver, British Columbia, Canada - The Vancouver Fire Department accepted their (A-9119) Pierreville (PTF-621) ladder unit on March 15, 1975, and placed the unit in service on May 20, 1975. It was the last mid-mount aerial to be ordered, and cost $79,325. The Imperial-supplied chassis, a 37,000-pound GVWR, has a Detroit Diesel 6V-71 motor coupled with an Allison automatic transmission. The aerial is a 100-foot Thibault mid mounted. A 250-gallon PTO booster pump is provided with a 200-gallon water tank. Little information is available on this unit other than the fact that it was sold in 1999. *Photo courtesy of Dan Goyer*

Spec. No. I–622 Vancouver, British Columbia, Canada - The Vancouver Fire Department took delivery of this Pierreville (PTF- 622) rear-mounted 4-section 100-foot Thibault ladder (A-9134 shop number) on May 3, 1975, on an Imperial custom fire chassis. The aerial apparatus has a 250-gpm booster pump and a 200-gallon water tank. It is powered by a 265 HP, Detroit Diesel 6V-71 motor with an Allison automatic transmission. The apparatus is one of two Pierreville (PTF-621) Imperial (I-621) ladder units purchased in the 1970s (the other was a midship unit). The rear-mount was sold at an auction on July 20, 1999. The mid-mount was also sold in 1999. *Photo courtesy of Dan Goyer*

Spec. No. I-623 Brick, NJ. - Herbertsville Fire Company Number 1, Brick Township District Number 3, received their Model D-10 in May 1974. A diesel, 8V-92 engine and an automatic transmission powered this custom truck. It featured a 1,000-gpm midship pump with a 500-gallon booster tank. High body side compartments, driver's side stainless steel pump panels, dual hose reels, and a top-mounted generator equipped the rig which served the district until May 2000 when the apparatus was sold to Montclair, New Jersey. A pre-owned Aerialscope was sold to Howell Township, New Jersey. The Imperial was traded and they purchased three American LaFrance Eagles. *Photo courtesy of Scott Mattson*

Spec. No. I-624 Dumont, NJ. - It is believed that the Dumont Volunteer Fire Company purchased the Model D-10 and it was delivered in April 1974. Sometime between then and February 28, 1994, it was traded in on an American LaFrance. They then sold it to one of their dealers, Utley's Fire Apparatus in Beebe, Arkansas. Southern Stone County Fire Protection District in Reeds Spring, Missouri added it to its fleet of three used Imperials. The rig, on a 180-inch wheelbase, is powered with a Detroit Diesel 8V-71, 318 HP motor and has an automatic transmission. The pump is a Waterous CMBX 1,000-gpm pump and it is mated with a 750-gallon water tank. *Photo courtesy of Jerry Bell*

Spec. No. I-625 Spring Mount, PA. - Lower Fredrick Fire Company placed their Imperial triple combination pumper Model D-10 in service in April 1974. Features include a Detroit Diesel 8V-71, Allison HT-740 automatic transmission, a Waterous CSMBX 1,000-gpm pump, a 750-gallon tank, two 1 1/2-inch pre-connect cross lays, one 1/2-inch pre-connect in hose bed, one 2 1/2-inch rear pre-connect pre-piped stationary deck gun, and a hose bed cover. NFPA equipment was included. The unit was sold in August 1999 to the Orangeburg County Fire District in Orangeburg, South Carolina. *Photo courtesy of Todd Lincoln*

Spec. No. I-626-2 Glenolden, PA. - Darby Township acquired three Imperial rigs that were delivered in June 1974 for Goodwill Fire Company (I-626-1) in Glenolden, traded to Ferrera and later sold to Overcamp Fire Department Number 1 in Morriton, Arkansas in February 1999. One for BriarcliffeVolunteer Fire Department (I-626-2) in Glenolden, sold on September 11, 1986, to Colcord, Oklahoma; one to Darby Township Volunteer Fire Department Number 4 (1-626-0) in Sharon Hill, salvaged in 1994 and is stored at the lot of Enforcement Towing. Chassis specifications were as follows: Ford C-900, 153-inch wheelbase, 534 engine, Spicer 6352 transmission, 9,000-pound front axle, 18,500-pound rear axle, 1,000-gpm pump and a 500-gallon tank. *Photo courtesy of Richard Walthall*

Spec. No. I-628 Levitown, PA. - Edgley Fire Company Number 1, Bristol Township placed this Model DA-4S aerial Quint in service in June 1975. A contract for $88,578 was signed on August 27, 1973. The 240-inch wheelbase, 8V-71, 350 HP engine, Allison HT-70, 4-section 100-foot LTI midship-mounted aerial ladder featured a CMBX 1,000-gpm two-stage pump, a 150-gallon water tank, rear-mounted booster reel, two 600-foot 2 1/2-inch hose beds, ladder intercom, 110-volt outlets at ladder tip, two 33-cubic foot air bottles with piping up the ladder, lifting bar at end of fly section, Ross dump valve, stainless steel pump panels, and more. The unit tentatively will be sold to Groesbeck, Texas in November 2002. *Photo courtesy of Greg Donaghy*

Spec. No. I-631 Audubon, NJ. - Defender Fire Company Number 1 placed their Imperial mid-mounted 100-foot LTI aerial in service in June 1975. The 46,000-pound GVWR chassis has a 6-71N engine, and an HT-750 transmission on a wheelbase of 230-inches. It carries 233-feet of ground ladders, 600 feet of 2 1/2-inch hose, and 200 feet of 3-inch line. Note the fully enclosed rear body. Defender Fire Company ceased fire protection in the fall of 1996, a new Audubon Fire Department was created, and the aerial unit was transferred to the new department. The Imperial aerial was replaced with an E-One aerial in February 2001 and sold to a private individual in Atco, New Jersey in the spring of 2002. *Photo courtesy of Dave Ehrman*

Spec. No. I-632 Beecher Falls, VT. - Beecher Falls Fire Department received their Pierreville (PTF-637) pumper in July 1974 on an Imperial chassis ordered in 1973. In January 1976 the fire station was destroyed by fire and the only truck that was salvaged was the Imperial, which was accomplished by pushing the tines of a forklift through the cab and dragging it from the burning station. It was sent back to Drummonville to be re-built and returned to service in the fall of 1976. In 1998 it was completely rebuilt including a poly tank, a wiring harness, an automatic transmission and paint, along with some other makeovers. It continues to serve the community today. *Photo courtesy of Ron Bodgardus*

Spec. No. I-633 Bensenville, IL. - The Bensenville Fire Department placed their Pierce Manufacturing 85-foot platform in service in October 1975. The aerial rig is mounted on a 60,500-pound GVWR Imperial chassis 234-inch wheelbase with a Detroit Diesel 8V-71, 350 HP engine, an Allison automatic HT-750 CRD transmission, a Rockwell 16,500-pound front axle and a 44,000-pound tandem rear axle. The midship pump is a Waterous 1,500-gpm, single-stage, CSUMBX with a 200-gallon tank. The platform was manufactured at Ladder Towers Inc. The rig served the department until February 1999 when it was sold to the Boardman Rural Fire Protection District. *Photo courtesy of Chuck Madderom*

Spec. No. I- 634 Colonial Beach, VA. - Colonial Beach Volunteer Fire Department took delivery of their Model D-7 triple combination pumper in January 1975. The 178-inch wheelbase 36,000-pound GVWR rig engine is a Detroit Diesel 8V-71N, and the transmission, an Allison HT-750. The Waterous pump is a CMHBX-750 3-stage, and the water tank is 1,000 gallons. Wheelhouse compartments, covered reels, permanent pre-piped deluge gun, polished aluminum wheels, bumper recessed Q siren, fire bell, rear canopy rail, and roof-mounted air horns (now illegal) adorn this beauty which, after a quarter of a century, still serves the fire department. *Photo courtesy of Skip Stinger*

Spec. No. I-635 Halifax, Nova Scotia, Canada - The Halifax Regional Fire & Emergency Service purchased a Pierreville triple combination pumper on an Imperial chassis and put the rig in service in February 1977. The 37,080-pound GVWR 190-inch chassis was manufactured in July 1976. Chassis features included an 8V-71N (328 HP) motor and an HT-740D automatic transmission. The body was completed by Pierreville and included a 2,000-gpm (1,750 Imperial Gallon) Waterous pump, and a 500-gallon booster tank. An electric horse reel for 200 feet of 1-inch hose is mounted on the left-hand side over the pump. The unit is still used extensively and is in reserve status today. *Photo courtesy of Don R. Snider*

Spec. No. I-638 East Rockaway, NY. - The East Rockaway Fire Department placed their Imperial Model D-15 low profile pumper in service in April 1975. Major components consisted of a Detroit Diesel 8V-71N engine, a Spicer syncromesh Model 6852 transmission, a Waterous 6852 1,500-gpm midship pump, and a 600-gallon booster tank. Value-added features included a front suction, cross lays, and over-the-wheelhouse compartments. Warning and lighting equipment included dual roof-mounted air horns, a recessed beacon ray light, amber and red warning lights, siren and bell. The unit was sold in April 1995 to the Nassau County Training Academy in Bethpage, New York. *Photo courtesy of Roland Boulet*

Spec. No. I-639 Woodbury, NJ. - Friendship Number 1 Fire Department received their Model DT-3 85-foot Ladder Tower in August 1975. The contract was signed on December 7, 1973, at a total cost of $113,137. It featured a 62,000-pound chassis with an 18,000-pound front axle and a 44,000-pound tandem rear axle, an 8V-71N, 350 HP engine, Allison HT-740 transmission, and a low profile 5-man cab. Ground ladders provided 180 feet. The unit was refurbished by Ladder Towers in 1993, and the cab was redone by Duplex (Duplex ceased operations in December 1997), and is in service today. *Photo courtesy of Bob Warfield*

Spec. No. I-640 Lubbock, TX. - The Lubbock Fire Department ordered two Pierce "Fire Marshall" pumpers that were placed on Imperial chassis (I-640/641) and delivered in April 1975. The 33,200-pound GVWR chassis, 178-inch wheelbase rigs were equipped with 12,000-pound front axles and 24,000-pound rear axles, Cummins power NTF-295 HP, Spicer syncromesh transmissions, Waterous CMBEX 1,250-gpm pumps, and 500-gallon water tanks. Dual electric hose reels and dual cross lays were provided. The units were sold at auction in 1995 for $5,200 to an individual who resold them to a volunteer department in Alabama. The disposition of the other unit is unknown. *Photo courtesy of Mike Hendricks*

Spec. No. I-647 Bensenville, IL. - Bensenville Fire Department took delivery of their Pierce "Fire Marshall" on an Imperial chassis in March 1975. The 36,000-pound GVWR chassis is powered with a Detroit Diesel 8V-71N engine, an Allison HT-740D automatic transmission, and is equipped with a Waterous CMBX 1,250-gpm midship pump and a 750-gallon water tank. Extra-value features include a front suction, a fire bell, a recessed 888 light, dual CP-25 roof-mounted speakers, twin beacon ray lights, dual pre-connected hose cross lines, high side compartments, and many other items. This custom apparatus remains in service with the department today. *Photo courtesy of William Freidrich*

Spec. No. I-648 Montreal, Quebec, Canada - The Montreal Fire Department acquired three Pierreville custom 1,250-gallon pumpers on Imperial chassis. Two units—I-648 (PTF-488) and I-650 (PTF-490)—were delivered on June 3, 1975, and unit I-649 (PTF-489) was delivered on June 4, 1975. The 36,000-pound GVWR chassis is powered with a Detroit Diesel 6L-71 motor, and features an Allison HT-740 automatic transmission. The tank capacity is 625 U.S. gallons. Unit PTF-488 was in service until December 16, 1996, unit PTF-490 remained in service until April 16, 1996; today their whereabouts are unknown. Unit PTF-489 was sold to the City of Charlemagne on December 22, 1975. *Photo courtesy of Pierre Gascon*

Spec. No. I-653 Kelowna, British Columbia, Canada - The Kelowna Fire Department placed their Pierreville aerial in service on August 6, 1976. The Thibault 4-section midship-mounted 100-foot aerial is mounted on a 40,000-pound GVWR custom Imperial chassis with a wheelbase of 190 inches. It is powered by a Detroit Diesel 265 HP with an Allison transmission Model HT-740, 16,000-pound front and 24,000-pound rear axles. Valued extra features include a Waterous Model CPK-2, Class A 300-gpm discharge rate, power steering, air service brakes, and 12-volt electrical system. The overall length is 39-feet, 3-inches, the overall height is 133 inches. It remains in service today. *Photo courtesy of Skip Stinger*

Spec. No. I-655 Whitby, Ontario, Canada - Whitby Fire Department took delivery of their top-mounted control panel Pierreville pumper in September 1976. The Imperial/Pemfab 40,000-pound chassis was completed in March 1975, with a 16,000-pound front axle and a 24,000-pound rear axle. The motor is a Detroit Diesel, the transmission an Allison HT-740, and it features a 1,000-gpm single-stage Waterous CSMB pump and a 500-gallon capacity water tank. The rig was sold in December 1990 to the Ville de Vaudreuil-Dorion in Quebec, Canada for $89,358.20 and the color was changed from the original yellow to white over red. *Photo courtesy of Gary Dinkel*

Spec. No. I-662 Evansville, IN. - The Evansville Fire Department placed their Pierce (8547-C) rear-mounted 104-foot aerial in service on June 1, 1976. The aerial was manufactured by LTI (Ladder Towers Inc. of Ephrata, Pennsylvania) and mounted on a 68,000-pound GVWR low-profile 1974 Imperial, 240-inch wheelbase chassis on March 10, 1975. Chassis components consisted of a Detroit Diesel 8V-71N engine, Allison HT-740 automatic transmission, a 24,000-pound front axle and a tandem 44,000-pound rear axle. The heavy compartmentalized unit was finished with a white ladder and canopy cab roof and a sunflower yellow lower half. *Photo courtesy of Gregory Stapleton*

Spec. No. I-675 Louisville, KY. - Pleasure Ridge Park Fire District took delivery of this Pierce Manufacturing (8565-C) Model "Fire Marshall" triple combination pumper on September 15, 1975. The 178-inch wheelbase pumper was built on an Imperial 35,000-pound fire chassis I-675. The motor is a Detroit Diesel 8V-71N, the transmission an Allison HT-740D, and it features a Rockwell FF-901 12,000-pound front axle and an R-140 23,000-pound rear axle. The midship Waterous pumping capacity is 1,250-gpm while the tank size is 500 gallons. It has many valued features including a front suction and black anodized pump panels. This yellow and white rig is in service today. *Photo courtesy of Gregory Stapleton*

Spec. No. I-697 Evansville, IN. - The Evansville Fire Department received this Pierce (8071-C) Model "Fire Marshall" pumper on August 3, 1976. The 178-inch wheelbase Imperial chassis housed an 8V-71N engine, an Allison HT-74 transmission, an 18,000-pound front axle, a 24,000-pound rear axle, a Waterous 1,250-gpm CMBX pump and a 500-gallon tank. This unit is representative of the four Pierce/Imperial pumpers purchased by the city. The first one (I-656) was purchased in 1974 with a 6V-71N engine. Three more were purchased in 1976: I-696 A (8840-C), I-696 B (8840-C) and I-697 (8071-C), all with 8V-71N engines. *Photo courtesy of Gregory Stapleton*

Spec. No. I-698 Port Moody, British Columbia, Canada - The Petro-Canada Burrnard Products Terminal Fire Department placed their Pierreville (PTF-501)/Imperial foam pumper in April 1975. Powered by an 8V-71N engine and HT-740 transmission, the Imperial chassis was outfitted with a 14,000-pound front axle and a 23,000-pound rear axle. The original sunflower yellow-finished rig had a 1,250-gpm Waterous CSUYB single pump and a 1,200-gallon foam tank. The unit was rehabbed in 1997 by Anderson Engineering and Chubb National Foam who installed a new 1,400-gallon poly tank, a new pump house, and a new pump. They then repainted the rig all red with a white safety stripe. *Photo courtesy of Bob Dubbert*

Spec. No. I-701 Westmount, Quebec, Canada - The Westmount Fire Department placed their pumper (from Thibault) on January 19, 1976. This rig features a 36,000-pound GVWR, 204-inch wheelbase chassis with a Detroit Diesel 6V-71N engine, an Allison HT-740D automatic transmission, an Imperial 1,050-gpm pump, and a 500-gallon tank. It remains in service today and is one of the last Imperial chassis produced. Pemberton Fabricators now promotes "Chassis by Pemfab," and in 1985 changed to "Pemfab Trucks," which would endure until July 1997 when it was sold to CECO-Taylor, a division of Chambersburg Engineering that ceased operations in July 2000. *Photo courtesy of Claude Deziel*

INDEX BY LOCATION

ABOUT THE AUTHOR

Richard J. Gergel

Did you ever hear the comment "curious as a cat?" If so, you can relate to why this book happened. Pemberton Fabricators employed me between 1985 and 1996, and during my tenure I heard many good comments regarding IMPERIAL Fire Apparatus, but I could never locate any of the company's records. (Former employees had told me that the records were non-existent, and no one knew how they were disposed of.) Throughout my ten-year period at Pemfab many independent fire salesmen kept my curiosity alive, so it was only natural, faced with time on my hands and my desire to preserve history, that the book would become a reality. I did not know the research would take nearly three years to complete. Using my computer skills and through the friendships I've built in my many years in the fire industry, I compiled a list of amateur fire buffs and photographers and started building and collecting information. I might add that without their help this book would not have happened. Once again, I would like to acknowledge each of the many individuals involved, and thank them for their contributions and friendships.

Employed by Ward LaFrance Truck Corporation in Elmira Heights, NY, from June 2, 1952, I served in various capacities from blueprint operator to the company's President. I was appointed President the first time on October 11, 1973, until the company was sold to North Street Associates on November 5, 1975. After being named Chairman of Ward LaFrance International, a worldwide sales and export company, I was then appointed President of Ward LaFrance Truck Corporation for the second time on November 25, 1977. After leaving to become Division Manager of Mack Fire Apparatus on March 10, 1980, in Allentown, Pennsylvania, I was separated from Mack Trucks and later joined PEMFAB Trucks in Rancocas, New Jersey in July 1985 as Vice President of Marketing and Sales. A career change in January 1996, I was named Vice President and General Manager of SEFAC, an American subsidiary of AFE, Immobile Mirror of Montrouge, France until retirement in 1997. While at Ward, I served in Engineering, designing and styling the "Firebrand" and "Ultravision" trucks, moving through Sales Engineering to Sales Manager, where I was responsible for developing a national sales dealer network and then to Vice-President of Sales, where I was credited with selling the first "lime yellow" apparatus. As President, I was able to negotiate an $18.4 million contract with the Royal Kingdom of Morocco. While serving as President, I designed and styled the "Patriot," a low-cost custom tilt cab model. A member of America's Young Presidents Club at 39, I also served in industry associations, FAMA (Fire Apparatus Manufacturers Association) and FEMSA (Fire and Emergency Manufacturers and Service Association) and am most currently a member of SPAAMFAA (Society for the Preservation and Appreciation of Antique Motor Fire Apparatus). In 2000, I co-authored a *Photo Archive* history book on Ward LaFrance Truck Corporation, published by Iconografix.

You too can help complete, maintain, and perpetuate the IMPERIAL Fire Apparatus Registry. Simply email "RJGergel@aol.com" or write to Richard Gergel, C/O SPAAMFAA, PO Box 2005, Syracuse, NY 13220 or write to Iconografix indicating you would like your letter (with proper postage provided) forwarded. Please indicate: the chassis serial number, date of manufacture, current owner, manufacturer's job number, model, your location, and your address.

MORE TITLES FROM ICONOGRAFIX

*This product is sold under license from Mack Trucks, Inc. Mack is a registered Trademark of Mack Trucks, Inc. All rights reserved.

All Iconografix books are available from direct mail specialty book dealers and bookstores worldwide, or can be ordered from the publisher. For book trade and distribution information or to add your name to our mailing list and receive a **FREE CATALOG** contact:

Iconografix, PO Box 446, Dept BK, Hudson, Wisconsin, 54016   Telephone: (715) 381-9755, (800) 289-3504 (USA), Fax: (715) 381-9756

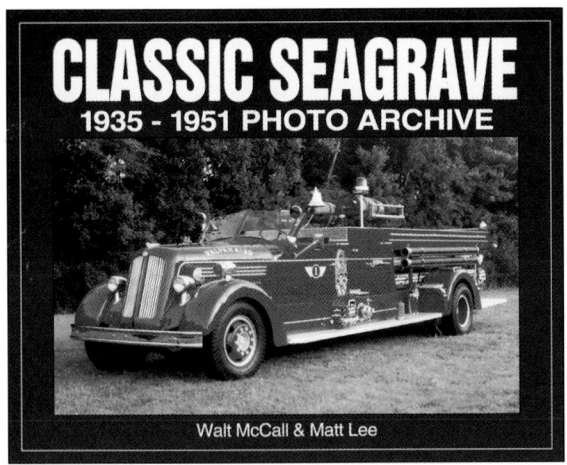

CLASSIC SEAGRAVE
1935 - 1951 PHOTO ARCHIVE
Walt McCall & Matt Lee

MORE
GREAT BOOKS
FROM ICONOGRAFIX

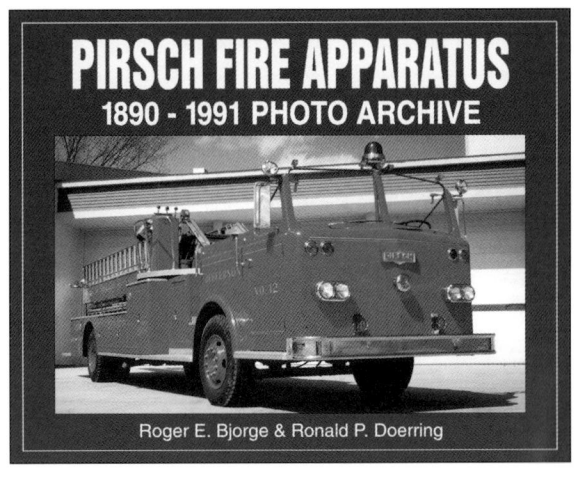

CROWN FIRECOACH
1951 - 1985 PHOTO ARCHIVE
Chuck Madderom

INDUSTRIAL & PRIVATE FIRE APPARATUS
1925 - 2001 PHOTO ARCHIVE
Leo E. Duliba

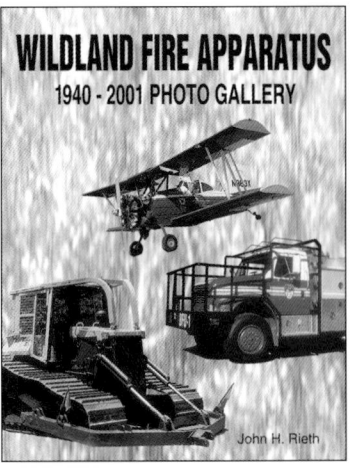

PIRSCH FIRE APPARATUS
1890 - 1991 PHOTO ARCHIVE
Roger E. Bjorge & Ronald P. Doerring

An Illustrated History
The American Ambulance
1900-2002
Walter M. P. McCall

HAHN FIRE APPARATUS
1923 - 1990 PHOTO ARCHIVE
Tom W. Shand

WILDLAND FIRE APPARATUS
1940 - 2001 PHOTO GALLERY
John H. Rieth